## abdobooks.com

Published by Abdo Zoom, a division of ABDO, P.O. Box 398166, Minneapolis, Minnesota 55439. Copyright © 2020 by Abdo Consulting Group, Inc. International copyrights reserved in all countries. No part of this book may be reproduced in any form without written permission from the publisher. Fly!™ is a trademark and logo of Abdo Zoom.

Printed in the United States of America, North Mankato, Minnesota.
102019
012020

Photo Credits: Alamy, Everett Collection, iStock, Shutterstock, ©Copeinator123 Amazing Fantasy Vol 1 15 p9 / CC-BY-SA
Production Contributors: Kenny Abdo, Jennie Forsberg, Grace Hansen
Design Contributors: Dorothy Toth, Neil Klinepier

**Library of Congress Control Number: 2019941306**

**Publisher's Cataloging-in-Publication Data**

Names: Abdo, Kenny, author.
Title: Spiders / by Kenny Abdo
Description: Minneapolis, Minnesota : Abdo Zoom, 2020 | Series: Superhero animals | Includes online resources and index.
Identifiers: ISBN 9781532129513 (lib. bdg.) | ISBN 9781098220495 (ebook) | ISBN 9781098220983 (Read-to-Me ebook)
Subjects: LCSH: Spiders--Juvenile literature. | Spiders--Behavior--Juvenile literature. | Arachnids--Juvenile literature. | Insects--Juvenile literature. | Zoology--Juvenile literature.
Classification: DDC 595.4--dc23

# TABLE OF CONTENTS

Spiders....................... 4

Origin Story.................. 8

Powers & Abilities........... 12

In Action..................... 18

Glossary ..................... 22

Online Resources ............. 23

Index ........................ 24

Spinning webs and catching thugs like flies, Spider-Man truly can do anything a spider can.

With eight legs and eyes, spiders creepy crawl all around the planet. Despite their off-putting image, they do more good than bad for the world.

Swinging onto page in 1962, Spider-Man made his **debut** appearance in *Amazing Fantasy* issue 15. Due to popular demand, Stan Lee and Steve Ditko created the **spin-off** comic series *The Amazing Spider-Man* in 1963.

Stan Lee said he wanted a hero that teens could relate to. Peter Parker was a high school outcast. No one suspected his **alter ego** to be Spider-Man.

# POWERS & ABILITIES

There are more than 45,000 known **species** of spiders. The world's biggest spider is the Goliath birdeater. It can grow nearly one foot wide. The world's smallest spider is the Patu. They are so small that 10 of them can fit on a pencil tip!

13

A spider's silk is five times stronger than steel. Scientists have failed to reproduce the strength and flexibility of spider silk.

All spiders spin silk. They use the silk to climb and wrap up **prey**. Some use silk to make webs.

Spiders listen to and feel the silk strands of their webs to find trapped **prey**.

Spiders are important in building a strong **ecosystem**. They eat harmful insects and pollinate plants. Spiders are a good source of food for mammals, birds, and fish.

Spider-Man has web-shooters on both wrists. He uses them to shoot web fluid to wrap up criminals. Spider-Man also uses them to swing from building to building, traveling through the city at high speeds!

Spider-Man has spider-senses. It is a tingling sensation that is connected to his reflexes. It helps Spider-Man avoid injuries and incoming danger.

Peter Parker has the **IQ** of a genius and is a skilled **acrobat**. He uses those skills only for the good of those around him. Rightfully earning his nickname, the friendly neighborhood Spider-Man!

# GLOSSARY

**acrobat** – a person who is highly skilled in gymnastics involving agility and balance.

**alter ego** – a person's secondary persona.

**debut** – a first appearance.

**ecosystem** – a community of organisms and their surroundings.

**IQ** – short for intelligence quotient. A score from several standardized tests to assess one's intelligence.

**prey** – animals hunted or killed by other animals for food.

**species** – living things that are very much alike.

**spin-off** – a new story pulled from an existing story that focuses on a particular character or event.

# ONLINE RESOURCES

**Booklinks**
**NONFICTION NETWORK**
**FREE!** ONLINE NONFICTION RESOURCES

To learn more about spiders, please visit abdobooklinks.com or scan this QR code. These links are routinely monitored and updated to provide the most current information available.

# INDEX

*Amazing Fantasy* (comic) 9

*Amazing Spider-Man, The* (comic) 9

Ditko, Steve 9

intelligence 21

Lee, Stan 9, 10

Parker, Peter (character) 10, 21

senses 16, 20

silk 14, 15, 16

species 12

Spider-Man 5, 10, 19, 20, 21

web 5, 15, 16, 19